A Trip Around the World

Using Expanded Notation to Represent Numbers

Kerri O'Donnell

Math
for the
REAL World™

Rosen Classroom Books & Materials
New York

Published in 2004 by The Rosen Publishing Group, Inc.
29 East 21st Street, New York, NY 10010

Book Design: Michael Tsanis

Photo Credits: Cover © SuperStock; p. 4 © ROB & SAS/Corbis; p. 7 © PhotoDisc; p. 8 (Fez) © Karen Huntt
Mason/Corbis; p. 8 (Sahara desert) © Andrea Pistolesi/The Image Bank; p. 11 © Aaron Strong/The Image
Bank; p. 12 © Harald Sund/Photographer's Choice; p. 15 © Travel Pix/Taxi; p. 16 © Walter Bibikow/Taxi;
p. 19 © V.C.L./Taxi; p. 20 © Lee Snider/Corbis.

Library of Congress Cataloging-in-Publication Data

O'Donnell, Kerri, 1972-
  A trip around the world / Kerri O'Donnell.
      v. cm. — (PowerMath)
Includes index.
Contents: A trip around the world — Off to London — An African
adventure — To the pyramids — The journey to India — Moscow, Russia
— Visiting China — The mountains of Peru — On to Canada — Home sweet
home!
  ISBN 0-8239-8966-6 (lib. bdg.)
  ISBN 0-8239-8871-6 (pbk.)
  6-pack ISBN 0-8239-7379-4
  1. Decimal system—Juvenile literature. 2. Voyages and
travels—Juvenile literature. [1. Decimal system. 2. Number systems. 3.
Voyages and travels.] I. Title. II. Series.
  QA141.35 .O36 2004
  513.5'5—dc21
                          2003001285

Manufactured in the United States of America

Contents

We couldn't wait to see how many miles we would travel on our dream vacation!

A Trip Around the World

If you could take a trip around the world, where would you go? There are so many interesting places in the world to see! My friend and I decided to plan the kind of trip we would want to take. We read about different countries around the world, then made a list of our favorite places and started planning our dream vacation. Our first stop would be London, England.

Off to London

To get a better idea about how far we would travel on our trip, we will use **expanded notation**. Expanded notation lets you look at every digit in a number to see which place value it belongs to, such as ten thousands, thousands, hundreds, tens, or ones. It is like taking a number and breaking it down into an addition problem.

Let's see how this works by using expanded notation to show how far we would travel on the first part of our trip to London. We live in Chicago, Illinois, so the trip to London would cover a distance of 3,949 miles (6,355 kilometers).

London is one of the world's largest cities. About 7 million people live there. London is the capital of the United Kingdom, which is made up of England, Wales, Scotland, and Northern Ireland.

England

London

Europe

3,949 miles =
3 (thousands) 9 (hundreds) 4 (tens) 9 (ones) or
3,949 miles = 3,000 + 900 + 40 + 9

6,355 kilometers =
6 (thousands) 3 (hundreds) 5 (tens) 5 (ones) or
6,355 kilometers = 6,000 + 300 + 50 + 5

Fez

Fez
Morocco

Africa

Sahara desert

Part of a desert called the Sahara lies in southern Morocco. The Sahara is the largest desert in the world!

An African Adventure

From London, we would travel 1,232 miles (1,982 kilometers) to Fez, a city in Morocco. Morocco is a country in northwestern Africa. It is bordered by the Atlantic Ocean to the west and the Mediterranean Sea to the north. These two bodies of water are connected by the **Strait** of Gibraltar. Where the Strait of Gibraltar is narrowest, only 8 miles (13 kilometers) separate Morocco and Spain. Morocco has many different kinds of land—plains, mountains, and desert.

trip distance in expanded notation:
1,232 miles = 1,000 + 200 + 30 + 2
1,982 kilometers = 1,000 + 900 + 80 + 2

To the Pyramids

The next part of our dream trip would take us 2,129 miles (3,426 kilometers) across the **continent** of Africa to a city called Cairo (KY-roh). Cairo is the capital of Egypt, a country in northeastern Africa. More people live in Cairo than in any other city in Africa.

Cairo is on the eastern side of the Nile River. At 4,145 miles (6,671 kilometers) long, the Nile is the longest river in the world! We both agreed we'd want to see the famous **pyramids** of Egypt, which were built about 4,500 years ago in the desert west of Cairo.

trip distance in expanded notation:
2,129 miles = 2,000 + 100 + 20 + 9
3,426 kilometers = 3,000 + 400 + 20 + 6

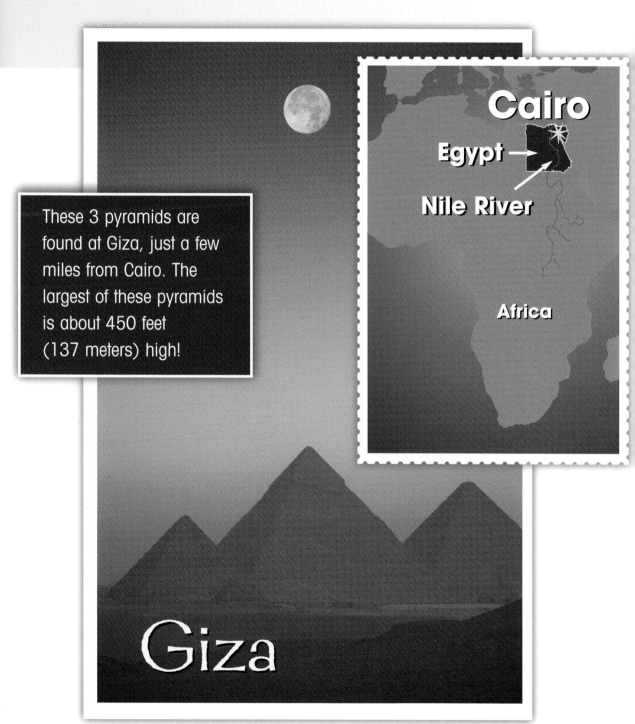

These 3 pyramids are found at Giza, just a few miles from Cairo. The largest of these pyramids is about 450 feet (137 meters) high!

Cairo

Egypt

Nile River

Africa

Giza

Agra

The center of the Taj Mahal is covered by a large, rounded ceiling that measures about 70 feet (21 meters) across and about 210 feet (64 meters) from the floor to the top.

Asia

Agra

India

The Journey to India

From Cairo, Egypt, we would travel 2,825 miles (4,546 kilometers) to a city called Agra in northern India to see a beautiful building called the Taj Mahal. A wealthy Indian ruler had the Taj Mahal built as a **tomb** for his wife, who died in 1629. It took about 20,000 men more than 20 years to build the Taj Mahal!

The Taj Mahal was built from white **marble**. It is one of the most famous and most visited places in all of India.

trip distance in expanded notation:
2,825 miles = 2,000 + 800 + 20 + 5
4,546 kilometers = 4,000 + 500 + 40 + 6

Moscow, Russia

The next part of our journey would take us 2,810 miles (4,522 kilometers) to the city of Moscow in western Russia. Moscow is the capital of Russia. Almost 9 million people live there, which makes it one of the world's largest cities. It was named after the Moscow River, which flows through the city.

Moscow is known for its interesting buildings. One of the most famous buildings in the world is St. Basil's **Cathedral**, a colorful church with onion-shaped **domes**.

trip distance in expanded notation:
2,810 miles = 2,000 + 800 + 10 + 0
4,522 kilometers = 4,000 + 500 + 20 + 2

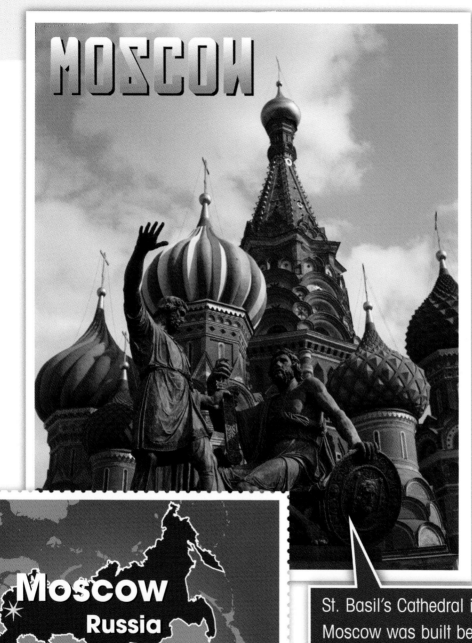

MOSCOW

Moscow
Russia

Asia

St. Basil's Cathedral in Moscow was built between 1555 and 1560.

GREAT WALL OF CHINA

Beijing

China

Asia

Part of the Great Wall near Beijing is about 35 feet (11 meters) high and 25 feet (7.6 meters) wide at its base.

Visiting China

From Moscow we would travel 3,601 miles (5,795 kilometers) to the city of Beijing (BAY-ZHING) in northeastern China. Beijing is China's capital, and more than 7 million people live there!

We decided we would visit the Great Wall of China while we were in Beijing. The Great Wall is the longest **structure** ever built—it stretches about 4,000 miles (6,437 kilometers) across China. Part of the Great Wall is found near Beijing.

trip distance in expanded notation:
3,601 miles = 3,000 + 600 + 1
5,795 kilometers = 5,000 + 700 + 90 + 5

The Mountains of Peru

Our next stop would be the **ruins** of the ancient city of Machu Picchu (MAH-choo PEEK-choo) in Peru, about 10,540 miles (16,962 kilometers) away! Peru is a country in western South America. A mountain range called the Andes runs north to south through the middle of Peru. Machu Picchu is located high up in the forests of these mountains.

The stone buildings at Machu Picchu were built by a native South American people called the Inca and were probably used by the families of the Inca rulers.

trip distance in expanded notation:
10,540 miles = 10,000 + 500 + 40 + 0
16,962 kilometers = 10,000 + 6,000 + 900 + 60 + 2

MACHU PICCHU

Peru

Machu
Picchu

South
America

The Inca had one of the largest empires in the Americas. The area ruled by the Inca included parts of what are now the countries of Peru, Ecuador, Chile, Bolivia, Colombia, Ecuador, and Argentina. Their empire was destroyed when the Spanish arrived in the 1530s.

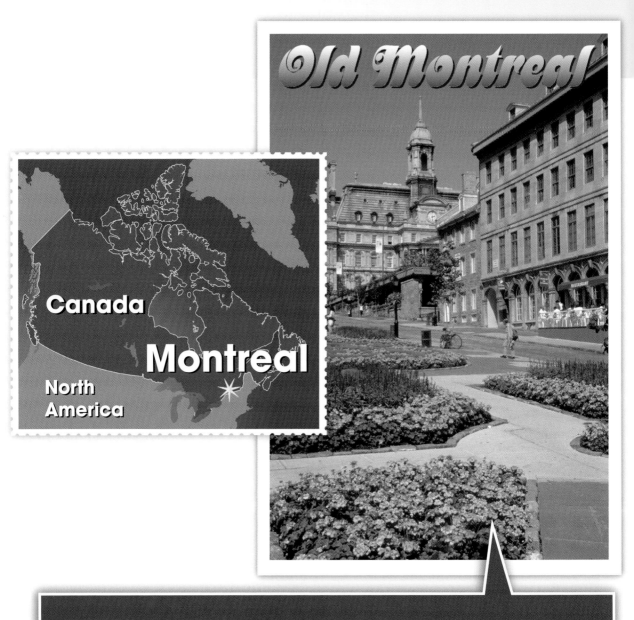

Old Montreal

Canada

Montreal

North America

In 1535, a French explorer named Jacques Cartier became the first European to reach the place that is now Montreal.

On to Canada

From Peru, we would travel 4,087 miles (6,576 kilometers) to Montreal, a city in eastern Canada. Montreal is the second largest French-speaking city in the world. Only Paris, France, is bigger. About $\frac{3}{4}$ of the people in Montreal speak French.

Montreal has one of the world's largest **inland** seaports. Part of Montreal's waterfront is called Old Montreal. In Old Montreal, old stone buildings line **cobblestone** streets. Visitors can see what the city looked like hundreds of years ago.

trip distance in expanded notation:
4,087 miles = 4,000 + 80 + 7
6,576 kilometers = 6,000 + 500 + 70 + 6

Home Sweet Home!

We decided that Montreal would be the last place we would visit. From Montreal, we would travel 738 miles (1,188 kilometers) back to Chicago to share our stories and pictures from the trip with our families and friends. Our dream trip covered a total distance of 31,911 miles (51,356 kilometers)!

total trip distance in expanded notation:
31,911 miles = 30,000 + 1,000 + 900 + 10 + 1
51,356 kilometers = 50,000 + 1,000 + 300 + 50 + 6

What places would you visit if you could take a trip around the world? Wherever you decide to go, you can use expanded notation to show how far you have traveled.

Glossary

cathedral (kuh-THEE-druhl) A large church.

cobblestone (KAH-buhl-stohn) A rounded stone used to pave streets.

continent (KAHN-tuhn-uhnt) One of Earth's 7 large landmasses.

dome (DOHM) A rounded roof that may look like half of a ball, an egg, or an onion.

expanded notation (ik-SPAN-ded noh-TAY-shun) A way of writing out large numbers as an equation that lets you see which place value every digit belongs to.

inland (IN-lund) Away from the coast.

marble (MAR-bul) A kind of hard, smooth rock often used in building.

pyramid (PEER-uh-mid) A building with a square base and 4 triangular sides that meet in a point at the top.

ruins (ROO-uhns) What is left after a building or structure has fallen apart.

strait (STRAYT) A narrow body of water that connects 2 larger bodies of water. The Strait of Gibraltar connects the Atlantic Ocean and the Mediterranean Sea.

structure (STRUHK-chur) Something that people have built. Houses, office buildings, churches, temples, dams, tombs, and pyramids are all structures.

tomb (TOOM) A large grave that is built above ground.

Index